Volcano Science
Volcanologists

by Julie Murray

Dash!
LEVELED READERS

Level 1 – Beginning
Short and simple sentences with familiar words or patterns for children who are beginning to understand how letters and sounds go together.

Level 2 – Emerging
Longer words and sentences with more complex language patterns for readers who are practicing common words and letter sounds.

Level 3 – Transitional
More developed language and vocabulary for readers who are becoming more independent.

abdobooks.com

Published by Abdo Zoom, a division of ABDO, PO Box 398166, Minneapolis, Minnesota 55439. Copyright © 2023 by Abdo Consulting Group, Inc. International copyrights reserved in all countries. No part of this book may be reproduced in any form without written permission from the publisher. Dash!™ is a trademark and logo of Abdo Zoom.

Printed in the United States of America, North Mankato, Minnesota.
052022
092022

Photo Credits: Alamy, Getty Images, Science Source, Shutterstock
Production Contributors: Kenny Abdo, Jennie Forsberg, Grace Hansen, John Hansen
Design Contributors: Candice Keimig, Neil Klinepier

Library of Congress Control Number: 2021950300

Publisher's Cataloging in Publication Data

Names: Murray, Julie, author.
Title: Volcanologists / by Julie Murray.
Description: Minneapolis, Minnesota : Abdo Zoom, 2023 | Series: Volcano science | Includes online resources and index.
Identifiers: ISBN 9781098228446 (lib. bdg.) | ISBN 9781098229283 (ebook) | ISBN 9781098229702 (Read-to-Me ebook)
Subjects: LCSH: Volcanoes--Juvenile literature. | Volcanologists--Juvenile literature. | Volcanism--Juvenile literature. | Physical geography--Juvenile literature.
Classification: DDC 551.21--dc23

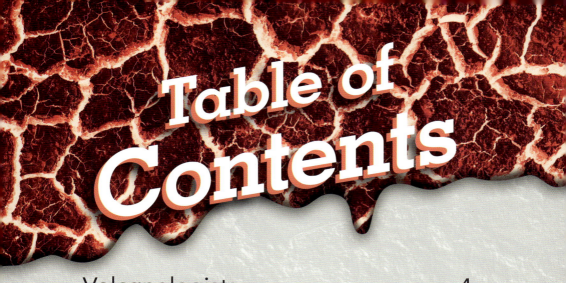

Table of Contents

Volcanologists 4

What They Do 8

Predicting Eruptions 18

More Volcano Facts 22

Glossary 23

Index 24

Online Resources 24

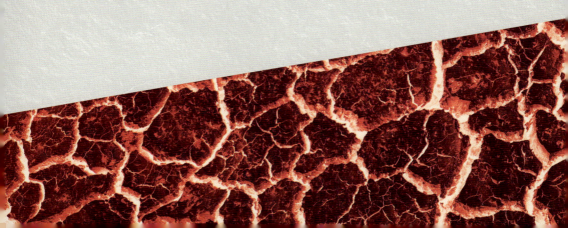

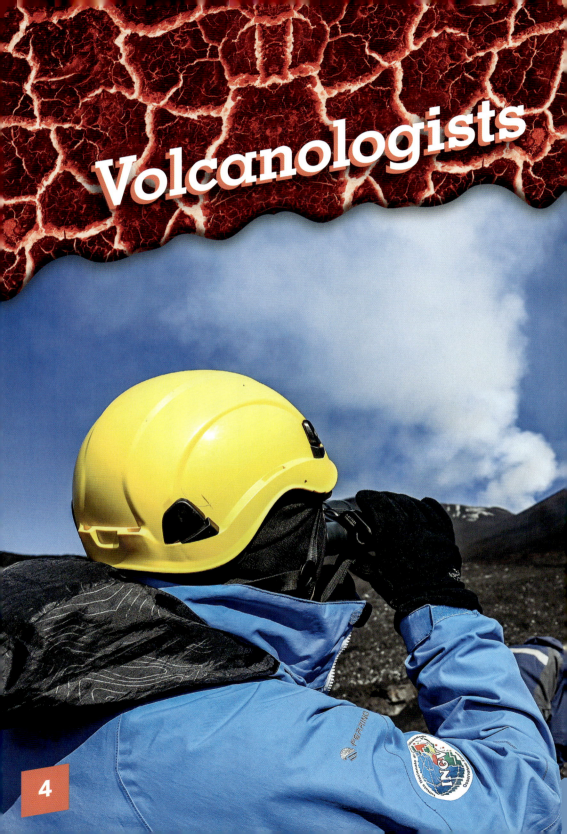

Volcanologists are scientists who study volcanoes. They study how and why volcanoes erupt. They try to **predict** when a volcano will erupt again. They also study the impact volcanoes have on the Earth and human life.

Their work is done in many ways. Some are on site collecting **samples**. Others are in labs studying the samples. Some use tools to record volcanic activity. Others work deep in the ocean!

What They Do

Volcanologists who collect **samples** work on site at the volcano. They collect ash, rocks, and lava.

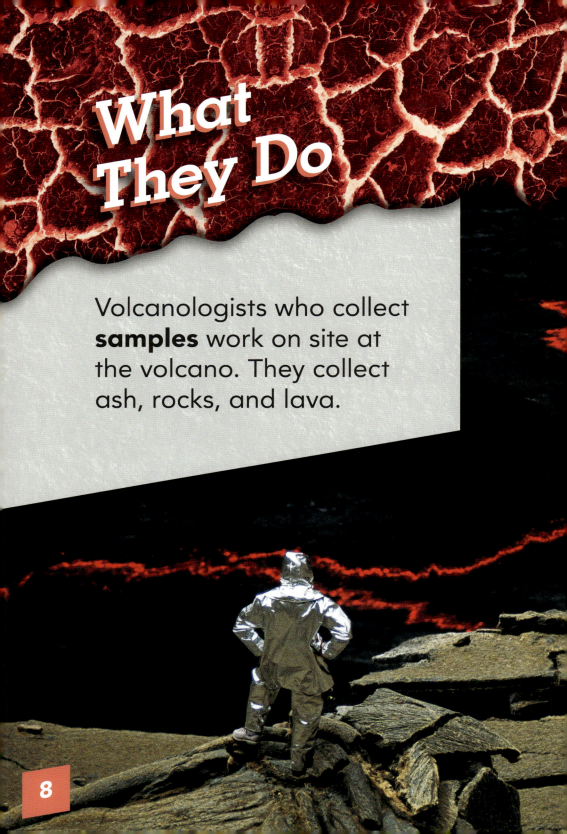

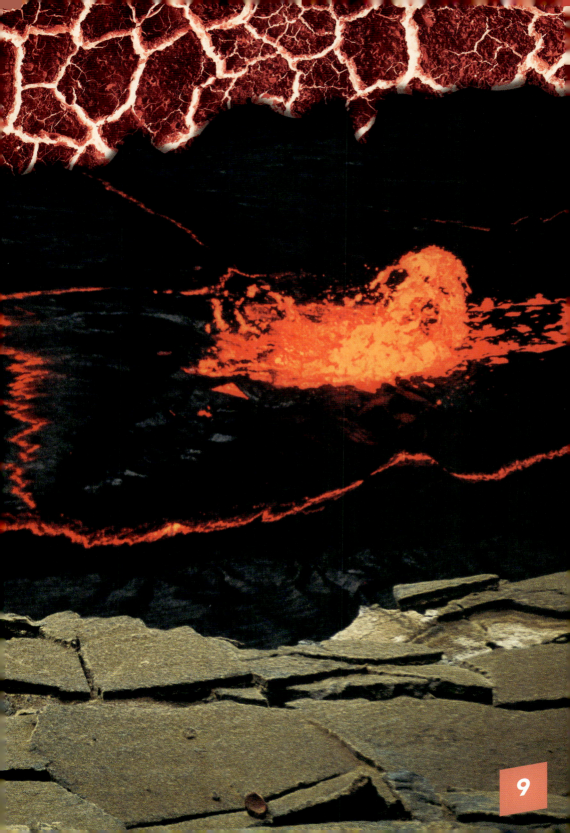

Scientists in labs study the **samples** taken at the volcano. They do research to learn more about volcanoes from the past and present.

Since many volcanoes are in the ocean, some volcanologists work under water. They **monitor** areas in the ocean that have volcanic activity. They also collect **samples** from these areas.

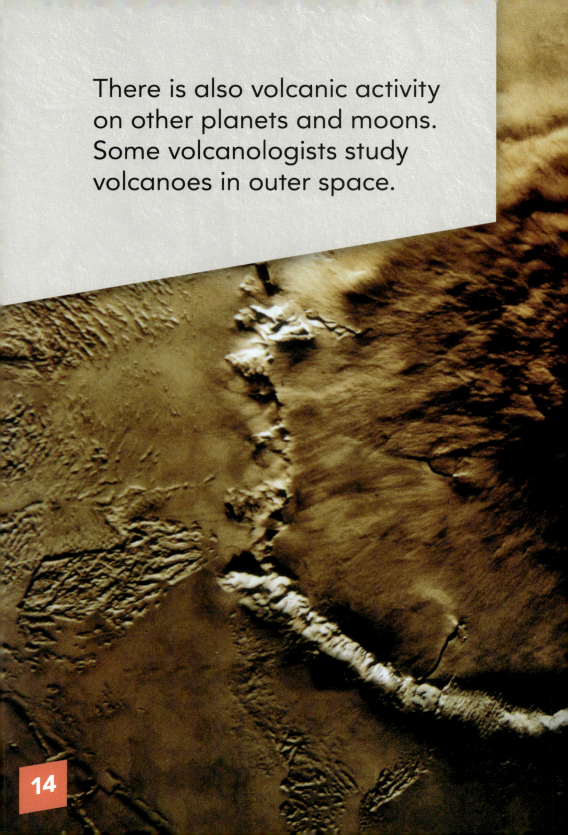

There is also volcanic activity on other planets and moons. Some volcanologists study volcanoes in outer space.

14

They work closely with the National Aeronautics and Space Administration (NASA) to collect **data**.

A volcanologist's job can be dangerous. David A. Johnston was a volcanologist. He died while working near Mount St. Helens when it erupted on May 18, 1980.

Predicting Eruptions

Volcanologists try to **predict** future eruptions. They study and observe activity at volcanic sites. They use drones and **satellites** to see changes in the volcano.

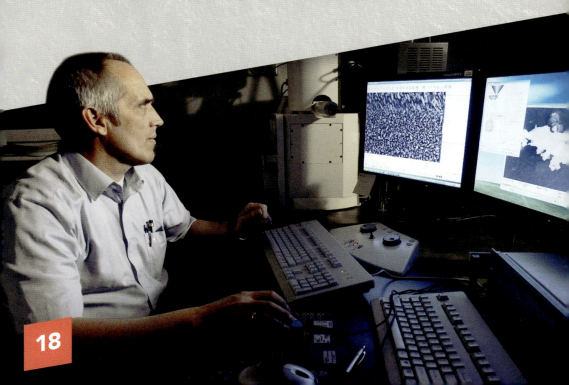

They also closely **monitor** earthquake activity in the area. This can be a sign of a pending eruption.

Volcanologists gather a lot of information. They alert people to **evacuate** an area if a volcano is going to erupt. Their work saves lives!

More Volcano Facts

- One of the first volcanologists was Pliny the Younger. He detailed the eruption of Mount Vesuvius in 79 CE.

- About 2,000 people study volcanoes.

- Since 1960, 31 volcanologists have died while on the job.

- Most volcanologists work for the government, universities, or private research companies.

- The United States Geological Survey (USGS) was formed in 1879 to **monitor** volcanic activity in the US.

Glossary

data – facts, figures, and other pieces of information that can be used to learn about something.

evacuate – to move or take away from a dangerous place.

monitor – to observe in order to check on.

predict – to tell in advance that something will happen.

sample – a small part of something that shows what the whole is like.

satellite – a spacecraft that is sent into orbit around a planet to gather or send back information.

Index

ash 8

dangers 16

duties 5, 6, 8, 11, 13, 14, 18, 19, 21

earthquakes 19

eruptions 16, 18, 19

Johnston, David Alexander 16

laboratories 6, 11

lava 8

Mount Saint Helens 16

NASA 15

ocean 6, 13

outer space 14, 15

tools 6, 18

Online Resources

Booklinks NONFICTION NETWORK
FREE! ONLINE NONFICTION RESOURCES

To learn more about volcanologists, please visit **abdobooklinks.com** or scan this QR code. These links are routinely monitored and updated to provide the most current information available.